CENTRAL OREGON
CAVES

by
Charlie and Jo Larson

ABC Publishing
13318 N.E. 12th Avenue
Vancouver, Washington 98685
(206) 573-0306

ISBN 1-886168-01-6

WARNING: *Caving is hazardous.* It is your responsib-
ility to get qualified instruction in safe caving, including
equipment, techniques, safety measures, and backup
systems. This book is sold with no liability to the author
or publisher, expressed or implied, in case of injury or
death to the purchaser or reader.

CONTENTS

CAVE MYTHS AND REALITY

CAVING IS HAZARDOUS: Caves, like mountains, obviously contain potential dangers but careful, properly equipped cavers who follow proven safety rules seldom run the risk of more than a bruise. Nearly all accidents in caves result from failure to apply common sense, safety rules, and/or a lack of suitable gear.

COLLAPSE: Caves do collapse for various natural reasons, but occurrences are widely separated in time and collapse-prone areas are easily spotted. The chances of being hit by natural rockfall are few; however, many cavers have been bruised by rocks carelessly dislodged by themselves or others.

BAD AIR: Pockets of gas seldom occur naturally in caves as in mines (a good reason to avoid mines), but gasses from volatile liquids (like gasoline) spilled in a cave can concentrate in low areas, creating the possibility of an explosion.

VALUABLE MINERALS: Cave minerals (sometimes called formations) deteriorate once removed from the cave, some rather quickly, and there is no market for them. The only marketable cave deposit is bat guano and there's precious little of that in central Oregon.

GETTING LOST: In central Oregon caves, it is virtually impossible for a reasonably calm adult with adequate lighting to get lost. (If you must, use removable route markers — reflective tape works well — and remove them on the way out.)

DANGEROUS ANIMALS: With one notable exception (below) "dangerous" animals are unknown in central Oregon caves. However, cave animals are wild and it is possible to contract disease from their nests or from the animal itself, so it is advisable to leave both alone. Bats are the most maligned of all "dangerous" cave creatures, having been the victims of centuries of bad press. They are certainly not blind and they do NOT fly into your hair. They are the only mammal to have attained true flight and their highly developed echolocation sense enables them to avoid all obstacles, even in total darkness. Like some other animals found in caves — mice, squirrels, racoons, badgers, rats, porcupines, etc. — bats are wild animals, may carry diseases, are apt to bite or scratch if captured, and should not be handled. If left alone, they are no threat to humans (They are a threat to insects.) and like other wild animals, do not deserve to be tormented.

RATTLESNAKES: Occasionally, in warmer, dryer parts of the state, rattlesnakes have been encountered at cave entrances where they move in and out to compensate for temperature variations or to prey upon other animals entering or leaving (including bats). Occasionaly they are trapped in pits or entrances with overhanging edges. They need be no threat if not surprised, cornered or threatened and will go their own way if allowed to. At some locations (like Lava Beds Nat. Monument) they are a protected species.

SPELUNKERS = CAVERS: Most people view caves with a certain amount of apprehension, perhaps even dread, but there are a few who visit them regularly. Their activity is often referred to as "spelunking" and they are often called "spelunkers" but they much prefer the names "caving" and "cavers." Virtually all caving is recreational in nature, but there are a few who visit and study caves for scientific purposes and are known as "speleologists."

TO PROTECT YOURSELF

Take more than one independent light source for each person in the party. Sharing a light will severly limit safe movement in the cave, could well make the visit a pain instead of a pleasure and greatly increases the chance of injury — not to mention the possibility that others may be depending on you for light! Experienced cavers recommend 3 light sources. Examples are: gas lantern, electric lantern, flashlight, miner's-type light (either electric or carbide and made to attach to a hardhat), chemical light (Cyalume, etc.), candles. Candles are the best of backup lights. They are cheap, compact and easily carried, almost indestructible, most reliable, and they provide warmth and cheer during long waits. The best combination is a miner's light securely mounted on a hardhat (leaves the hands free), a powerful flashlight to highlight cave features and for backup, a chemical light or candles. Flares or torches are poor lights. They are somewhat dangerous, they flicker unreliably and they emit smoke (which dirties the cave) and gases which can be lethal to cave life (including you). They are also very annoying to others. For easy caves (no crawling) a Coleman-type lantern and a flashlight make a good combination. Appropriate spare parts for lights are important: lantern mantles and fuel, bulbs, batteries and especially matches which are vital to ignite some of the above lights.

Head protection is of the utmost importance. It is virtually inevitable that heads will collide with projecting parts of the cave and, if unprotected, a scalp cut or worse is often the result. Also, loose rocks can be dislodged by others. A hardhat, preferably with chin strap is strongly recommended. Even a cloth cap — though not adequate — is better than nothing.

Go slowly during your first few minutes inside the cave to allow your eyes time to adjust to the darkness. Also, the entrance zone is the most likely place to find ice which can be treacherous on inclined floors and sometimes all but invisible. If a trail is apparent, use it. Unstable rocks are less likely where others have passed before.

Avoid risky situations. Stay back from pits, sharp drops, water-filled passages and unstable areas unless you are equipped for such obstacles. Don't attempt anything that may be beyond the capabilities of the least able member of your party. Running, jumping and climbing ropes hand-over-hand are risky. Avoid using equipment found in a cave — it is likely to be deteriorated. Exploring a cave alone is risky. In the event you are injured or stranded without light, it would be far longer before you are found than if companions could have gone for help. A minimum of three people is recommended.

Dress appropriately. Cave temperatures in central Oregon vary with altitude and local climate. They are generally in the 35° to 50° range but some are colder, much colder, if they are cold air traps or at high altitude. Also, cave humidity is relatively high which, when combined with low temperature makes hypothermia (exposure sickness) a genuine threat. Dress warmly in several layers to permit shedding clothing to accommodate various levels of exercise. (Wool is recommend for cold or wet caves.) Lavatubes are hard on shoes. Sneakers are barely better than nothing. Boots with lug soles are recommended. Crampons, boot chains or a hand line are advisable for steep ice slopes. *A slide down a steep ice slope is virtually the equivalent of a free fall of the same height.*

COMFORT ITEMS: Items which can contribute greatly to the enjoyment of caving, but which aren't regarded by some as necessary include: gloves, knee pads, compact foods and water if your stay in the cave is long (don't drink cave water), first aid kit, and two trash bags: one to use as a raincoat if the cave is dripping water, or to conserve body heat during periods of inactivity, the other to carry litter in or perhaps for clothes dirtied in the cave. Finally, a small backpack is very handy for carrying everything because, as with the head-mounted light, it leaves the hands free. If, for some unforeseen reason, your party becomes stranded in a cave, stop, make yourself comfortable and wait for help which will arrive **IF** you have taken the following precaution:

BEFORE ENTERING

Leave word with someone, or somewhere, regarding your whereabouts and when you expect to return. Otherwise, should the entire party become stranded in a cave, it would be extremely difficult if not impossible for rescuers to find you. At minimum, consider leaving a note on or in your car, or at your residence. Remember that all caves belong to somebody and if there is any uncertainty regarding a cave's status — especially if a gate is found ajar or a signpost without a sign is found nearby — then seek out the owner or managing agency for clarification and permission to enter.

TO PROTECT THE CAVE

Caves are uncommon geologic features. Some cave features (formations) are even rarer; they aren't found in all caves and are seldom found in abundance. It is important to protect them in their natural setting for the benefit of others. Do not deface a cave with writing, marking or painting, or by using torches or flares. Campfires in caves are literally life threatening; to cave life as well as the visitor's life. If you wish to mark your way, use temporary markers (reflective tape works well) and remove them on your way out. Try to minimize foot and hand prints. If a trail through the cave is obvious, use it.

Carry out everything that's carried in and, if possible, remove trash left by other less considerate visitors. Try to leave every natural thing as it was. Personal sanitation requirements should be met outside. Don't disturb cave formations in any way. They are scarce and grow (if at all) with infinite slowness. There is no market for them (in many states it is illegal to collect or trade in cave minerals), and they are not good collectibles as they quickly lose their attractiveness once removed from the cave. Most will quickly dry out and become dull, and some actually disintegrate outside. Even if formations are found broken, leave them lay, for their presence outside the cave might encourage others to collect. Do all your collecting with a camera.

ANTIQUITIES

Many central Oregon caves and adjacent areas are archaeological sites and are protected from disturbance by the Federal Antiquities Act of 1906, and state laws forbidding excavation, collection or destruction of artifacts. The remains of prehistoric and historic cultures belong to all of us. When artifacts are stolen and archaeological sites destroyed, we lose important clues about the past, forever. Oregon law (ORS 358.905) forbids excavation or collection without a permit, sale or purchase of artifacts, and destruction of archaeological sites. Permits to excavate are issued only to recognized scientific or educational institutions. If artifacts are found, they are best left alone and reported to the property owner or land managing agency, as penalties under the above laws are severe.

WHY ALL THE PRECAUTIONS?

1. There is an element of danger in visiting caves, statistically less than mountain climbing, for example, but more than, say, fishing.

2. Caves are a finite natural resource, few more will be created in any reasonable time frame, and if their quality is to endure for the benefit of future generations, they need to be well treated.

It is the author's belief that the overwhelming majority of people, once acquainted with what's to gain and what's to lose, and the relative risks and dangers, will behave responsibly, and it is in that spirit that the foregoing tips on protecting yourself and the cave are offered.

Wooden ramp built by turn-of-the-century ice miners at Arnold Ice Cave. From a postcard in the Silver collection.

INTRODUCTION

In the western U.S., there are two principal types of caves: caves in limestone (dissolved by water) and caves in basalt, left by the drainage of lava and called either lava caves or lavatubes, plus several lesser types. Limestone caves are found in broad belts which may span several states; for example, in the Klamath Mountains of southern Oregon/northern California, or the Wallowa Mountains of northeast Oregon/western Idaho.

Most central Oregon caves are lavatubes and most of those are located on the flanks of Newberry Volcano, the largest in the 48 states. Most are very much older than they appear, on a human scale, but very young on a geologic scale. The ages of some of the volcanic landforms in central Oregon are known rather precisely, for example, Lava Butte, 6,000 years; but ages of lavatubes are largely educated estimates. All of the Newberry lavatubes described herein are at least 6,800 years old, for they contain ash from the eruption of Mount Mazama. Caves in the Cascades, like Sawyers Caves, have no such ash and are perhaps as young as 3,000 years, perhaps younger as the most recent vulcanism there is guessed to have been about 1,000 B.C. The presence of Pleistocene bones in Skeleton Cave indicates greater age. Some of the caves, like Sawyers, look young. They have "rough edges," while others are clearly altered by erosion, a time-consuming process indicating much greater age. This diversity is, of course, to be expected in a region where vulcanism has been and will continue to be on the menu. If not buried by subsequent eruptions, lavatubes erode quickly, geologically speaking.

Ice forms in nearly all central Oregon caves each year but lasts year around in only a few. It is created when water enters and encounters sub-freezing surroundings. Cold air, being heavy, settles in caves. Some caves collect more cold air than others, sometimes so much that the rock surrounding them is chilled to sub-freezing temperature to a depth of several feet. Basalt does not exchange heat readily and if the cave is poorly ventilated (warmer air can't get in because there is no air movement), the walls may remain cold well into spring, or even year around as in Arnold Ice Cave. The best time to see cave ice is late April to early May.

Central Oregon caves are perhaps the most culturally rich in the western U.S. Human utilization of them began many thousands of years ago and has been more or less continuous to the present. Except for some parts of the Horse Lavatube System, in and around the present city of Bend, the caves aren't habitable year around; however, all were valuable as seasonal hunting camps and sources of water (from ice) and some appear to have been used for ceremonial purposes. Later, pioneers and settlers put them to good use as well: for cold storage and summer ice, the latter providing ice cream as a centerpiece of many a weekend picnic. South Ice Cave and Derrick Cave were popular with residents of the Fort Rock and Christmas valleys and the road from La Pine to South Ice Cave was once known as the "Ice Cave Road." Ice was commercially mined from Arnold Ice Cave for many years prior to widespread availablity of refrigerators and repeal of prohibition, which sharply reduced the practical value of it and other ice caves. A few caves and open trenches continued to be used as stock corrals and some still are. Skelton, Pictograph and several other caves in and around Bend are known to have harbored distilleries.

During World War II, many central Oregon caves were considered as possible shelters for war plants, later as bomb shelters, still later as fallout shelters. One was even stocked with provisions as a fallout shelter (Derrick Cave). Today, use is almost entirely recreational in nature. Occasionally they are used for educational purposes; school outings, OMSI tours, etc. Scientific use is minimal. Two caves are commercialized: Lava River and Lavacicle; both are publicly owned and operated.

HOW LAVATUBES ARE FORMED

LAVATUBES are roofed-over segments of lava rivers. They may be occupied by lava (active), plugged with solidified lava (basalt), or empty. If empty, and having an entrance, they may qualify as a cave; i.e., a natural subterranean cavity large enough for a person to enter.

Lava rivers form only in lava which is very fluid when erupted, a basaltic type called Pahoehoe (a Hawaiian word pronounced Pa hoy hoy). It has a high gas content and behaves somewhat like water; spreading in all directions if erupted onto a level surface, or downhill on a slope. If such an eruption is constant and lengthy, outer parts of the expanding lava field are immobilized by loss of velocity, heat and dissolved gases — or are confined by topography — and channels form. The channels become rivers and, like rivers of water, can meander, overflow their banks, build levees, erode downward, and will freeze if their temperature drops. Depending on the length and steadiness of eruption, lava rivers may carry lava for days, months, even years. They are a primary mechanism for spreading lava over wide areas and their tendency to overflow their banks is important to the formation of deep lavatubes like Lava River Cave and the Arnold System. They are literally built from the ground up as successive overflows deepen the channel. The ridges built by sideways overflow, though readily seen on topographic maps, are not obvious to one standing on the ground because of their great width and gentle slope. *See diagram on next page —*

Under favorable conditions, crusts may form on the river's surface, in somewhat the way that ice forms on a river of water, beginning at the edges and growing toward the middle. If the river level remains stable, further cooling strengthens the crust until it is strong enough to support its own weight, even though the level of the river on which it floats drops away. A roof may be swept away by surges, or burst open by hydraulic pressure, only to form and reform again; or it may be strengthened by the addition of layers inside (from surges) or outside (from overflows). A roofed-over lava river becomes an even more efficient carrier of lava because less heat is lost to the atmosphere, and it can grow longer.

LAVATUBE (LAVA RIVER) SYSTEMS

Lava rivers are commonly composed of open channels, roofed-over sections (lavatubes), tumulii, and ponds too wide to roof over. When such a river's source "dries up" (eruption of lava stops), it drains, to the extent that its shape allows. Drainage is accompanied by decompression, cooling, contraction and failure of features too weak or plastic to stand unsupported by lava. At this time, much roof collapse occurs, shortening lavatube segments, opening "skylights" and creating collapse trenches. Finally at rest, the "dried up" lava river is a sinuous assembly of ponds, unroofed channels, collapse trenches, tumulii, and lavatubes (emptied or not), and is referred to as a lavatube *system*. Empty lavatubes are, of course, the component of most interest to cave explorers.

At this time, the system and its parts are in "original" condition. Unlike cave systems in limestone, lavatube systems cannot grow once the eruption that produced them stops. If not re-occupied or buried by a subsequent eruption, they begin a process of deterioration due to erosion. (All three of the above appear to have happened to the Arnold Lavatube System.)

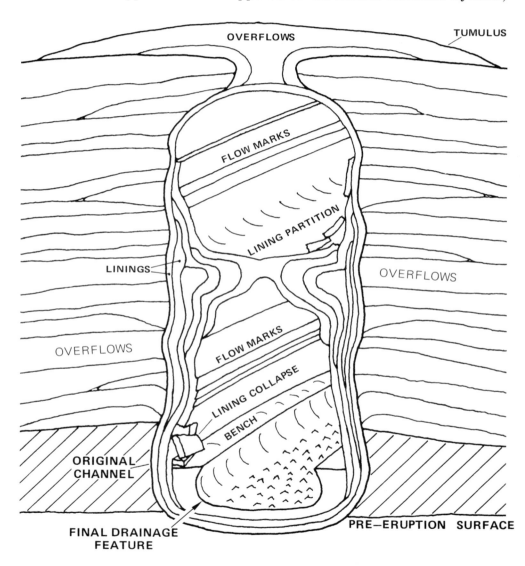

CROSS-SECTION OF A TYPICAL MULTI-LEVEL LAVATUBE: Note that unless collapse occurs, structural features are hidden from view, and scarcely any features result from final drainage. (The reader is cautioned not to expect ever to see a lavatube *exactly* like this.)

FEATURES OF LAVATUBES

The character and features of a lavatube result from conditions so complex, interacting and chaotic, that it is impossible to predict by looking at one section what will be found much beyond a few hundred feet away. Features can be analyzed, however, at specific points, which are best illustrated by cross-section. Keep in mind that most features result from the passage of lava past that point (primarily variations and interruptions of it). Few result from final drainage.

Often, a lavatube's history (or the lava river's history) is one of repeated draining and filling, each episode accompanied by deposition of a lining (see diagram). Thin linings (layers of hardened lava on existing walls) are deposited by abrupt surges or rapid passage. Prolonged or sluggish surges deposit thicker layers. If the level remains constant long enough, crusts form, growing from each wall toward the center and eventually joining to make a "lining partition" which, if the lava beneath drains, creates a multi-level passage. Variations produce hourglass, keyhole, minaret and skull shaped cross-sections. The number of shapes possible as a result of variations in the rate and duration of lava flow through a lavatube is staggering. One aspect of the buildup of linings that confounds understanding them is that the final lining (left as the last lava drained away) hides many, if not all, of the earlier linings. It is only at points of collapse, where the structure of the wall is revealed, that flow events can be studied.

ENTRANCES: Lavatube entrances are nearly always through a gap in the roof, which may be either an original or subsequent collapse, or simply a point where a roof didn't form. After a few thousand years of weathering, it is difficult or impossible to tell the difference.

CRACKS: Contraction cracks are an original feature, created as the basalt finally cooled. If it cooled rapidly, they are small, numerous and not obvious; slow cooling produces fewer but wider cracks. As a general rule, cracks account for about 1% of the surface area of a tube's interior. In central Oregon, it is through these cracks that most of the sand enters the caves.

OTHER TYPES OF VOLCANIC CAVES

In addition to lavatube caves, the predominant type in central Oregon, there are two other significant types:

SURFACE TUBES are small drained rivulets, or runners of the same highly fluid lava that flows in rivers. Formed initially on an existing hardened surface, they are seldom large enough to enter. They advance, literally, by turning themselves inside-out. Sometimes called "toes," they are thought by some to be instrumental in the growth (lengthwise) of lavatubes. They often form when lava vents, rivers or reservoirs overflow, and just as often are buried by further overflows. Examples of surface tubes are often seen in the walls of collapse trenches.

OPEN VERTICAL CONDUITS are vertical passages through which lava rose to the surface then receded. Their mouth is usually, though not necessarily, at the top of a vent structure like a spatter cone or spatter ridge. Hornitos form atop lavatubes. Refer to Skylight Cave for examples of hornitos.

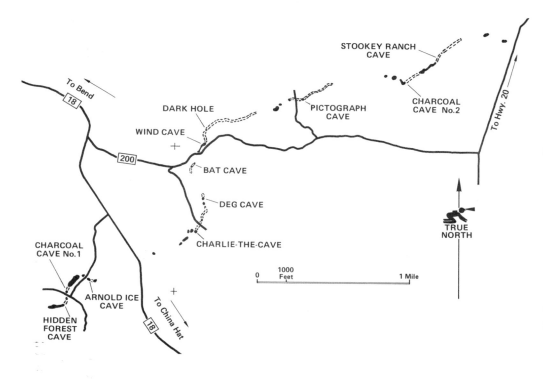

ARNOLD LAVATUBE SYSTEM

The Arnold Lavatube System is what remains of a great river of lava that flowed northwest from somewhere in the Kelsey Butte/Lava Top Butte area, the vent now buried by younger basalt. It has been traced for 3.9 miles (map length) and would be about 4.5 miles long if stretched out straight. It is an uncommonly deep multi-level system; three caves have lower levels about 135 feet below the surface. Several of the caves have overburdens in excess of 100 feet, unusual for lavatubes.

Ten of the cavernous segments have been named, some several times. Arnold Ice Cave, after which the system was named, has been renamed more times than any other cave in Oregon. Despite the many names, all are part of a common lava river, a unitary system which had, as far as is known, no tributaries or branches. Until 1972 the only known multi-level cave was Bat Cave No. 1, but in that year Stookey Ranch Cave was reported to have another level. Since then another, Charlie-the-Cave has been added to the list. There are several sizable unknown sections; for example, between Arnold Ice and Charlie-the-Cave, and between Pictograph and Charcoal No. 2. It is highly likely these sections are cavernous but entrances are not known to exist.

In the late '60s, a team from NASA studied several northwest lavatube systems in connection with the Apollo program. Results of their study of the Arnold System were published as Bulletin 71 of the Oregon State Dept. of Geology & Mineral Industries, *Geology of Selected Lava Tubes in the Bend Area, Oregon*. That publication is highly recommended to anyone wishing to delve deeper into the system's geology.

ARNOLD ICE CAVE

Arnold Ice Cave is one of a cluster of short caves and sinks — at the upper end of the Arnold System — referred to in the past as Crook County Ice Caves. The location is well marked on all maps of the Bend area and there are road signs leading there from all major roads and highways. Arnold Ice Cave's entrance is a relatively small opening at the base of a northwest facing cliff about 40 feet high, in a large, deep sink.

The accessible part of the cave is now only about 125 feet long and slopes down steeply to what is now barely recognizable as a small cupola, about 95 feet below the surface. Only near the bottom are features of the original tube present. **NOTE:** Negotiating this steep ice slope is extremely hazardous without crampons and/or a handline.

Arnold Ice Cave to the left of the turnaround. Charcoal Cave is at top, just below aircraft wingtip.

MORE CAVE

There are several accounts of more cave to the east now isolated by ice. In 1911, two large "bins," (probably rooms) " . . . about 15 feet across," were described. In the '50s, Bend cavers Jim Anderson and Phil Coyner chopped a trench in the ice, gaining access to nearly half a mile of passage, in turn blocked by ice. It is probable that the extension was a lower level, passing beneath the sink to the east, as multiple levels of comparable depth occur elsewhere in the Arnold System.

ICE

Arnold is the only cave in Oregon, possibly in the U.S., in which perennial ice is known to be advancing. It is not necessarily accumulating because the cave is completely plugged with ice, preventing winter air from reaching the now-isolated passage. There is a lot of cave beyond, no doubt of it, but whether its former ice is intact beyond the present plug is not known.

Arnold has all the requirements for the natural formation and preservation of ice. It has poor ventilation through a relatively small entrance, a steep slope to the bottom and more than adequate overburden (40 feet at the entrance, 95 feet at the bottom). It is coupled to a large sink (over 1,000 square yards area) which very effectively collects cold air (and water), and the entrance faces north. In one respect, Arnold differs markedly from other central Oregon ice caves — most of its recent perennial ice is formed of runoff water from the sink. Consequently, most of the ice is, to some degree, cloudy due to impurities — some layers having the appearance of frozen mud. It is the author's opinion that the recent ice advance is simply part of a long-term phenomena, halted briefly by the ice mining. It is likely that

View from the surface down into entrance of Arnold Ice Cave in 1967, about four years after construction of the stairway.

Arnold's ice forming capacity has undergone a series of temporary self-limitations — configurations in which cold air intake was strangled by ice accumulations, thus causing the ice to recede. Indeed, such a configuration may be developing now as the ice surface at the head of the buried stairway nears the ceiling.

HISTORY OF USE

There is evidence of human use of the Arnold caves as early as 1370 A.D., the carbon date from nearby Charcoal Cave. There is now strong evidence of human use of a cave only a few miles away, as early as 10,000 years ago. It is doubtful that Arnold Ice itself was ever used as shelter, however, because it remains at or below freezing year around. It was certainly used as a source of water and probably for cold storage, while the warmer Hidden Forest Cave nearby may have been used for shelter.

Ice miners posing on large blocks of ice quarried inside "Joe Cave" (Arnold Ice Cave). These blocks were hauled to the surface by horses, then loaded onto wagons for transport to Bend. This scene is just outside the entrance. One can only hope that the photographer was working as fast as possible. From a postcard in the Silver collection.

14

The earliest literature regarding Bend area caves is an 1889 *Oregonian* article about the "Crook County Ice Caves," referring to the Arnold Ice cluster. It was a favorite of settlers in the Bend area around the turn of the century and even gained international fame following its appearance on a number of picture postcards printed in Germany. Arnold was very popular with the loggers who worked the area during the '20s, for cold storage and recreation. The logging railroad (now the main road, No. 18) passed the ice cave about 1/4 mile to the northeast. One oldtimer who ran a skidder, and who claims to be one of the first to discover Wind Cave, recalls that a logger's cookhouse was about 100 yards from the Arnold sink. He recalls that various individuals kept their whiskey cool inside the cave.

Until the advent of refrigeration, much of Bend's ice supply was collected along the Deschutes River in winter and stored in sawdust-insulated buildings for summer use. This was not, however, a reliable source, as occasionally little ice was found. A tale of the happenings during one of the ice-lean years is a fascinating account of what may have been Oregon's only commercial ice mine. One year, an enterprising saloon keeper named Hugh O'Kane cornered the city's scant ice supply and hiked the price to an unheard-of 10 cents per pound. This created an immediate handicap for the two other saloon owners, which they solved by quickly surveying and building a road to the "Ice Cave," where they established an ice mining operation. To secure their new-found source, a cohort filed a homestead claim including the cave. The ice cave road of the '20s was quite different from today's. It went first to Skeleton Cave then went directly to Arnold where it circled around the cluster of caves.

In 1963, the Forest Service, attempting to improve the cave, installed a wooden stairway so that visitors could safely negotiate the steep ice slope. In less than ten years, rising ice had rendered the stairway virtually useless, even dangerous. On the other hand, it's certain that, for a few years, it provided access for many visitors, as well as preventing the chopping of steps in the ice, a practice that is once again gaining favor since the stairway became useless.

Sometime during the July Fourth weekend, 1975, the most destructive act of vandalism ever to occur at a central Oregon cave nearly closed Arnold. Vandals dislodged boulders from the cliff over the entrance, dangerously destabilizing the cliff face. The loosened boulders seriously damaged what was left of the stairway and nearly plugged the entrance. Despite repairs, the entrance remains unstable and accumulation of ice has made the interior hazardous. It would be prudent to consult the Forest Service in Bend about conditions at Arnold Ice Cave before entering.

NAME

Arnold Ice Cave has had more names than any other central Oregon Cave. Its present name came about through the accidental association of two road signs, one for the "Ice Cave" and one for the Arnold's homestead, both pointing down the same road. It has been known as the Largest of the Crook County Ice Caves, The Ice Cave, Crook County Ice Cave, Joe Cave, Ice Farm, Father Cave, and as "Arnold after the discoverer," which seems unlikely.

BAT CAVE No.1

Bat Cave No. 1 is one of the multiple-level segments of the Arnold System. It has two entrances which are easily found by following the collapse trench southwest from the main entrance to Wind Cave, crossing the road and continuing in the same direction for about 200 feet to the "skylight" entrance. About 140 feet farther (same direction), is a collapse sink, the second entrance. The easily negotiated skylight is the most direct and a far easier way in than the low passage between it and the sink.

From the skylight, the upper level leads northeast about 90 feet to a junction where collapse has joined the two levels. There is a 10-foot overhang which can be avoided by traversing along the left wall. About 30 feet beyond the northeast end of the junction room, up and over a steep breakdown slope, there is a small complex of levels connected by a small pit. The lower level, entered at the junction room, underlies the upper level leading south.

Looking south into lower level of Bat Cave No.1. The floor in the foreground is essentially original. Beyond, breakdown and sand accumulations have nearly blocked the passage. Shards of lining stand on edge along each wall. The minaret shape of the ceiling suggest an upper level which, of course, there is.

Except for the sand, coming in through cracks, this level is in near original condition. Of special interest are the wall scrolls, formed as final linings sagged and peeled away from the wall. The lower level is a cold air trap and near-freezing temperature there is not uncommon even as late as August.

Bat Cave No.1 was named Bat Cave by an Oregon Speleolgical Survey party which counted "14" bats there in 1961. (The "No.1" was added later to distinguish it from several other "Bat" caves in Oregon.) In the early '60s the Oregon State Board of Health, in the mistaken belief that Oregon faced a bat/rabies problem, initiated a survey of the state's bats. Bat Cave No.1 and other Bend area caves were some of the first places visited in the bat search.

Actually, there were relatively few bats in the caves. Only a small minority of the bats in the northwest hang around in caves, but that was not generally known then. Many of the cave bats were dispatched so that their brain tissue could be examined for rabies. None were found to be infected. Nevertheless, a bat banding program* was initiated and as recently as 1973, Bat Cave No. 1 was posted as a research site.

---

* Bat banding, a now discredited technique, consisted of attaching a metal band to a bat's wing so that its habits and movements could be traced. It was implicit that the banded bat, as well as many others in the process, would have to be disturbed repeatedly. In fact, the more times the bat could be relocated, the better.

16

CHARCOAL CAVE No.1

In 1928, Walter J. Perry, a Forest Service ranger and a persistent investigator of Bend area caves, discovered a large amount of charcoal in what would be named Charcoal Cave and later Charcoal Cave No.1 to distinguish it from other of the area's caves containing prehistoric charcoal. In 1938, Perry persuaded Dr. Luther Cressman, for years the principal investigator of early man in Oregon, to examine the site. Amid the charcoal they found partly burned pine marked by stone axes, subsequently growth ring dated to 1,370 A.D. To this day the reason for such large fires in the caves remains a matter of speculation. Many lean toward the notion

The roof of Charcoal Cave No.1 is very thin and a tribute to the stability of basalt.

they had some spiritual significance. Others theorized that the fires were for melting ice as a source of water, but Dr. Cressman pointed out that the latter idea was a bit far fetched since there was a fine spring only a few miles away.

HIDDEN FOREST CAVE

Hidden Forest Cave is the westernmost segment of the Arnold System so far identified. It is entered through a majestically impressive collapse trench so deep that only the tops of large trees which grow in it can be seen from a distance. The cave itself is short but wide and high and partially floored with sand. Ventilation is good, as it connects via a passage through breakdown with an unamed sink between it and Charcoal Cave. The Hidden Forest sink is about 70 feet wide, 60 feet deep and 600-700 feet long. Its walls provide a cross-section of the structure of a tube-fed lava flow. The overflows and buried surface tubes are clearly exposed. This is the most habitable of the segments of the Arnold System; ventilated enough to dispel cold night air and facing west.

Hidden Forest Cave on left, unamed sink center, Charcoal Cave No.1 in upper right.

Deg Cave.

DEG CAVE

Deg Cave is much like the upper levels of Charlie-the-Cave and Bat Cave. It is about 400 feet long, trending northeast from just beyond a column of collapse which cut through all levels of the system. The entrance is in one of two sinks connected by a narrow bridge which shows well the effect of plastic deformation; it settled before completely cooling. Deg is mostly collapsed but a few original features remain near the end. A shallow depression midway between Deg and Bat caves hints of a so-far-unknown section of the system.

CHARLIE-THE-CAVE

Charlie-the-Cave was known for many years to be only a small, shallow segment of the Arnold System between Arnold Ice Cave and Deg Cave. About 150 feet long, it was all breakdown. In the late '70s, a party from Bend excavated a blowing hole in the entrance sink and opened another significant multi-level part of the system. This new section is about 1,200 feet long, but its two levels bring the total passage length to around 2,400 feet. Neither of the two recently discovered passages are on the same level as the original cave, leaving the tanalizing prospect of an additional 1,000 feet or more of passage without any known entrance. Charlie-the-Cave underlies the chain of sinks between its entrance and Deg Cave. Much of the upper level and part of the lower level is in original condition. All levels are terminated by a column of collapse just short of Deg Cave. The downtube end of the upper level is not easily reached: the only known entry is through a hole in the ceiling of the lower level and requires vertical gear.

NOTE: The excavated entrance is unstable and further collapse could occur at any time. There hasn't been much traffic in the newly-discovered sections and consequently there is unstable breakdown. *This is a hazardous cave.*

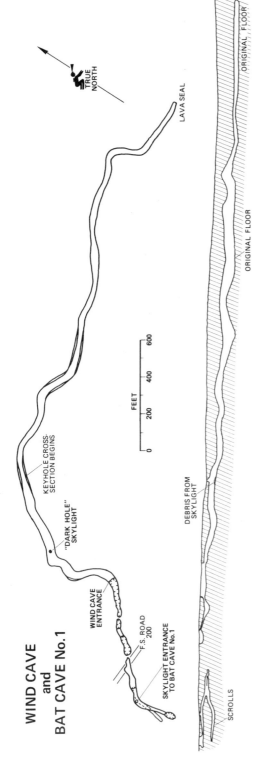

**WIND CAVE
and
BAT CAVE No. 1**

TRUE NORTH

LAVA SEAL

KEYHOLE CROSS-SECTION BEGINS

"DARK HOLE" SKYLIGHT

WIND CAVE ENTRANCE

F.S. ROAD 200

SKYLIGHT ENTRANCE TO BAT CAVE No.1

FEET

0 200 400 600

ORIGINAL FLOOR

ORIGINAL FLOOR

DEBRIS FROM SKYLIGHT

SCROLLS

Dark Hole.

WIND CAVE

Wind Cave is the largest segment of the Arnold System. At 2,700 feet (map length) it is about 1,100 feet longer than the next longest, Pictograph Cave. Its largest cross-sections are wider and higher than any other known cave in the system; and it is *rugged.* Anyone making the round trip to the end will easily understand why it was overestimated as a 5,000-foot-long rockpile by a caver who paced its length. The developed length (if stretched out straight and flat) is about 3,200 feet, the first two-thirds up and down over massive piles of breakdown. The back part, about one-third its length, is mostly original and it ends as the ceiling curves down to meet the floor behind a breakdown raft.

About 500 feet beyond the entrance, collapse has progressed upward to create a small skylight widely known as "Dark Hole." This skylight became a legend following a 1924 *Oregonian* story describing a cave entrance "..emitting an icy blast so strong as to carry light articles thrown therein aloft." Also known as Wind Hole, it is probably the origin of Wind Cave's name. Seasonal ice forms beneath the Dark Hole.

Interestingly, the 1930 Smithsonian report, primarily devoted to the bones found in Skeleton Cave, included an uncaptioned photo of the main entrance to Wind Cave.

PICTOGRAPH CAVE

Pictograph has been described as a short version of Wind Cave with spectacular entrances. The main entrance is through an off-center collapse with a steep slope. This sideways collapse provides a rare glimpse of what's behind the walls of a lavatube. Uncollapsed parts of the cave are large like Wind Cave's, up to 40 feet wide by 45 feet high midway between the entrance and the skylight, and have the same characteristic keyhole shape. The skylight is spectacular and a tribute to the strength of basalt cave roofs.

At one time, the cave was known as Stout Cave, after one of a pair of men who ran across it in 1957. At that time, it still housed a complete but rusty distillery. Briefly it was known as Charcoal Cave No. 2, following discovery of more ancient charcoal. The passage ceiling east of the main entrance is blackened by smoke of ancient fires. The cliff-like walls of the main entrance used to be a favorite nesting place of ravens, but as visitors increased, they have gone elsewhere.

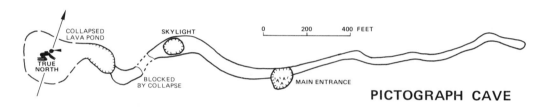

PICTOGRAPH CAVE

The skylight entrance to Pictograph Cave reveals a remarkably wide, thin roof. The wall of recently-fallen rock resting under the drip line shows how frost wedging is attacking the overhang.

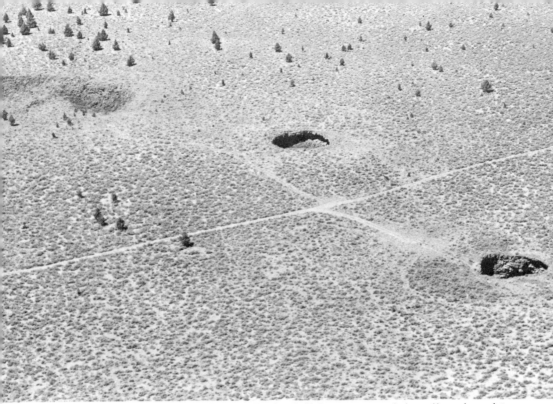

Aerial view of the main entrance of Pictograph, the skylight entrance and the collapsed pond to the west. While very obvious from the air, on the surface these entrances are very difficult to spot from a short distance away.

CHARCOAL CAVE NO. 2

Charcoal Cave No. 2 is the next east of Pictograph. Its impressive entrance leads to a steep slope down to a sandy floor. The only known passage is quite deep; multiple levels are not known to exist. In 1963, another unexplained deposit of charcoal was found in this cave. Because of interchangeable use of names and mistaken identities, there is confusion over which and how many of the system's caves contain it. Relatively large deposits have been found in Charcoal Caves No. 1 and No. 2 and in Pictograph. All observers agree on one point: large quantities of wood were carried or dumped into the caves when burned.

STOOKEY RANCH CAVE

Stookey Ranch Cave marks the eastern end of the Arnold System. Its multiple levels are reported to total about 4,000 feet of passage. In the first few hundred feet, two levels appear to be joined by collapse and there is a lot of dangerously unstable breakdown. Sand drains in at various points and may even have closed the entry passage near its end. Not far inside the entrance, on the left, a small passage leads up to connect with several small entrances in a small sink. Entrance to the main passage is not immediately obvious because of a ceiling offset. Stookey Ranch Cave is the third known multilevel part of the Arnold System. (The other two are Charlie-the-Cave and Bat Cave No. 1.)

21

Entrance to Boyd Cave.

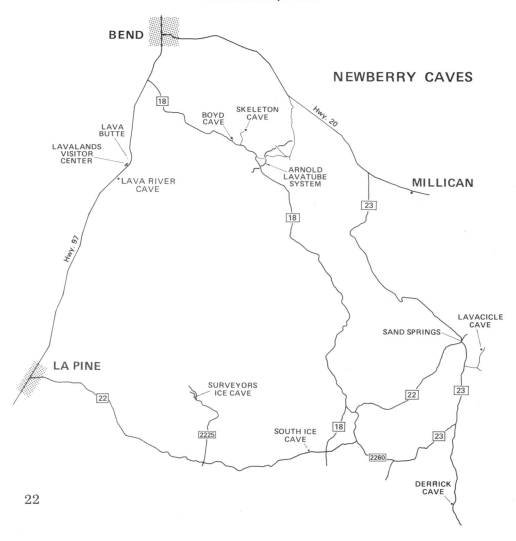

NEWBERRY CAVES

BEND

MILLICAN

LA PINE

18

Hwy. 20

Hwy. 97

BOYD CAVE

SKELETON CAVE

LAVA BUTTE

LAVALANDS VISITOR CENTER

LAVA RIVER CAVE

ARNOLD LAVATUBE SYSTEM

23

18

LAVACICLE CAVE

SAND SPRINGS

SURVEYORS ICE CAVE

22

22

23

2225

SOUTH ICE CAVE

18

2260

23

DERRICK CAVE

22

BOYD CAVE

Boyd Cave is only a short way north of the main Forest Service Road 18 about 1/2 mile southwest of Skeleton Cave. It appears to be about the same age as Skeleton but is shorter (about 1,880 feet compared to Skeleton's 3,000), and its roof is much thinner. Like Skeleton, the only entrance created short up-tube and long down-tube segments. A considerable pile of debris has accumulated beneath the small entrance, which overhangs all around, and probably includes the remains of many logs and ladders used before the Forest Service built a stairway in 1969. In the '70s, a very sturdy steel railing was installed around the entrance.

Boyd Cave's roof is so thin that plant roots penetrate into the cave and large quantities of sand are trickling in through contraction joints at several points. Collapse is minimal and many original features remain. Small alcoves are present all along the tube, which may be either cut banks or branches which failed to drain. About two-thirds of the way to the way to the end, partial drainage left a solid basalt crawlway about one foot high, beyond which walking passage continues. For many years, it was known in caving circles as Coyote Butte Cave, but the name Boyd was certified by a Forest Service publication around 1970.

The entrance to Boyd Cave is relatively small and the stairway barely fits.

DERRICK CAVE

Derrick Cave was part of a river of lava which began at a vent at the northeast corner of the Devil's Garden, a rugged 45-square-mile lava field, on the northern edge of the Fort Rock Valley. It is a section of that system through which much, perhaps most of the lava of Devil's Garden passed. Here, in a conveniently compressed form, may be seen the lava vent, an open lava river flowing away from it, short sections of roofed channel, and finally complete encasement of the lava river in a 1,200-foot lavatube.

The several entrances to Derrick probably resulted from collapse that occurred at final drainage. The area from the Main Entrance to a little beyond the skylights has a sand floor, lots of light (for a cave) and a peculiar beauty. This part of the cave is often used as a campsite and there's usually a fresh firepit there.

SOUTHEAST END

Beyond the skylights, the tube is essentially unmodified and quite different. It is relatively narrow and deep as roof thickness gradually increases to about 55 feet. Breakdown is moderate and many original features remain, notably, the tube-in-tube and leveed structures beginning just beyond the Big Room, and a long section of inflated floor near the end. The Big Room (46 feet high x 80 feet wide) is the point of departure for a high-level branch off the main tube, leading to Ben's Cave. **NOTE:** Ben's Cave has an overhanging entrance 12 feet deep. vertical gear is required.

NORTHEAST END

Northeast of the Main Entrance is a 600-foot tangle of levels, short unroofed segments and locally total collapse. Overflow structures, especially surface tubes, are plentiful. There are numerous smaller, shallower tubes each side of this area which may be overflow structures or may have originated at the vent as did the main tube.

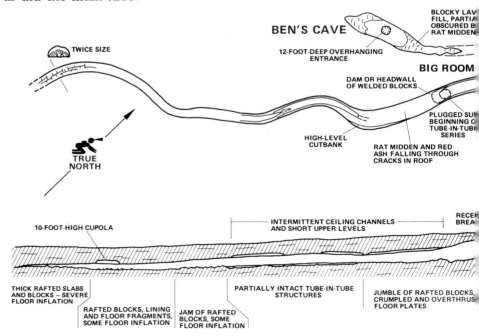

Light shining through the skylights of Derrick Cave creates the effect of another world.

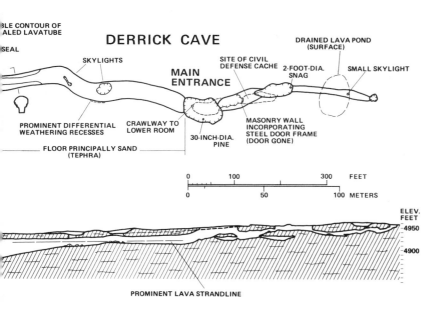

DERRICK CAVE

BLE CONTOUR OF
ALED LAVATUBE

SEAL

SKYLIGHTS

MAIN
ENTRANCE

SITE OF CIVIL
DEFENSE CACHE

DRAINED LAVA POND
(SURFACE)

2-FOOT-DIA.
SNAG

SMALL SKYLIGHT

PROMINENT DIFFERENTIAL
WEATHERING RECESSES

CRAWLWAY TO
LOWER ROOM

30-INCH-DIA.
PINE

MASONRY WALL
INCORPORATING
STEEL DOOR FRAME
(DOOR GONE)

FLOOR PRINCIPALLY SAND
(TEPHRA)

0   100   300   FEET

0   50   100 METERS

ELEV.
FEET

4950

4900

PROMINENT LAVA STRANDLINE

Derrick has been used by just about everybody, and abused by some. Early settlers in Fort Rock and Christmas valleys often picnicked there, sometimes 200-strong. A great attraction was the cave ice which was used to make ice cream, a relatively unusual treat for struggling homesteaders. Occasionally, ice persists year-around but Derrick is not listed as an ice cave.

In the early '60s, two projects were undertaken which were to leave their mark on the cave for years to come. The first was a joint effort by the State, Pacific Northwest Bell and North American Aviation, to determine if subterranean cavities and other phenomena could be located from far overhead. Their goal was to develop a method of locating and assessing possible havens beneath the moon's surface which Apollo astronauts might make use of. Two tons of railroad rail were lugged into the cave and loosely stacked at the end of the sandy floor about 200 feet beyond the last skylight. Then an attempt was made to locate the steel mass with instruments in an airplane furnished by the State and PNB. Later, the test was repeated after the steel had been heated with "gasoline" fires. The rails and smudgepots were simply left in the cave, but over the years, persons unknown have carried them off.

Steel doors, shown here in 1969, were installed at the entrance to the lower level. The locks were smashed off within weeks and now even the doors are gone.

Project No. 2:
In early 1963, civil defense agencies proposed to stock certain of the area's caves with supplies for use in the event of nuclear war. The project died for lack of funds, but not before one cave —Derrick—had been so prepared by the Lake County Search and Rescue Unit. A lower level was fortified with masonry and a steel door, then stocked with food, water and rudimentary sanitary facilities for 1,200 people. An eventual capacity for 3,000 was projected. The cache was quickly decimated by vandals operating with impunity in so remote an area, and today the steel doorframe remains but the door has been removed with a cutting torch. (The food went almost immediately. By 1970, nothing remained but the water, primarily because it was stored in steel barrels which had frozen solid.)

Derrick was named after H.E. Derrick, a pioneer rancher in the Fort Rock area, whose ranch was about 3 miles south-southeast of the cave and was known earlier as "Derrick's" Cave. Along with South Ice Cave to the northwest, it was a very popular recreation site for homesteaders in the valley. Also, it may be the elusive "East Ice Cave," so often referred to in literature but never positively identified.

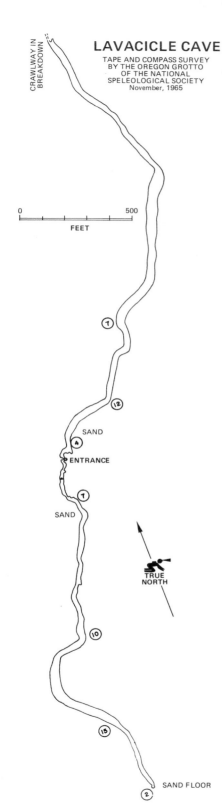

LAVACICLE CAVE

TAPE AND COMPASS SURVEY
BY THE OREGON GROTTO
OF THE NATIONAL
SPELEOLOGICAL SOCIETY
November, 1965

CRAWLWAY IN BREAKDOWN

0 _____ 500
FEET

⑦

⑫

SAND

Ⓐ

ENTRANCE

⑦

SAND

TRUE NORTH

⑩

⑮

SAND FLOOR

②

The Horses's Head, the largest drip-formed stalagmite in Lavacicle Cave was destroyed by vandals in 1979.

LAVACICLE CAVE

Lavacicle Cave has a map length of about 3,500 feet (4,231 feet traverse length). The only entrance divides it into a 2,429-foot north end and a 1,802-foot south end (traverse lengths). It reaches a maximum width of 60 feet in the north end, and a maximum ceiling height of 15 feet in the south end. It is believed to be about 8,000 years old. Unlike the typical lavatube, its entrance appears to have developed quite recently, perhaps thousands of years after the last lava drained away.

It is characterized by broad, smoothly arched cross-sections and jagged though easily traveled lava floors. The floor is covered locally with fine, light brown sand — washed in through the entrance and cracks — which blocks the passageway at the south end. Massive breakdown is locally prominent and creates some confusing passageways, especially at the entrance where the effect is complicated by the sand. The breakdown has created an exceptional opportunity to study the interior of a tube-containing flow. Most of the original

27

features remain. The ceiling and walls are mostly glazed and there isn't much to indicate hesitation in the lava flow. Apparently, much of the tube was completely full at the time of final drainage.

Considerable areas of the ceiling and walls are covered with a white precipitate, probably silica, which is most obvious at points of high air movement. It varies in shape from a hard, knobby coralloid to a soft hair-like coating. Where air movement is restricted — where evaporation is slight — the material has a scaly shape somewhat resembling small rimstone pools. Post flow conditions in Lavacicle were such that considerable lava was extruded from behind the tube lining. Most of it detached and fell to the floor where, depending on its temperature and the ambient temperature, it formed lava roses (cups), stalagmites or loose nodules. Some of the larger stalagmites (now vandalized or stolen) were up to four feet high.

Lavacicle Cave was discovered in August 1959, by a firefighting crew mopping up after the 23,000-acre Aspen Flat fire. Dan Beougher, a crew foreman noted a column of clean air rising through the smoke and found a small vertical entrance. Three crew members entered the cave down through a small offset passage. They were probably the first humans to enter the cave and the profusion of lava formations they encountered has become legendary. Not long after the discovery, the cave was visited by District Ranger Henry Tonseth and Deschutes National Forest Supervisor, A. A. Proust. According to Proust, the cave was to be named Plot Butte Cave, after the nearest butte to the south.

The name didn't stick. Metropolitan newspapers insisted on calling it "Pilot Butte Cave," probably mistakenly after the prominent butte in downtown Bend. Finally, about a year after its christening as Plot Butte Cave, Phil Brogan, writing for the Bend Bulletin, suggested it be named Lavacicle Cave, after the impressive lava formations it contained.* In late 1964, the U.S. Board of Geographic Names approved the new name.

The first experienced cavers to visit the cave were Jim Anderson, Phil Coyner and Joe Stenkamp of Bend. They found a considerable amount of water near the (north) end, a condition never since noted. They also found the skeleton of a small mammal later identified as a river otter, which created somewhat of a puzzle since the nearest body of water likely to sustain such an animal was about 25 miles away.

Fortunately, the Forest Service recognized the exceptional nature of the site and gated it at once. Many are the gates that have suffered since and many are the lava formations damaged or no longer there. Nevertheless, the Forest Service should be complimented on their handling of a difficult situation — a situation made no easier by a full page article in a Portland paper about the "Hidden Fairyland Beneath the Oregon Desert."

Lavacicle Cave is designated as a Geologic Site. The entrance is gated and can only be entered by appointment. To join a tour (usually on the third Saturday each month during tourist season) contact Lava Lands Visitor Center, (503) 593-2421.

---

*The earliest known use of the term "lavacicle" is in "The Lava River Tunnel," by Ira A. Williams, which appeared in Natural History in 1923.

The characteristic "skull" shape of well-developed lavatubes may be seen at several places in Lava River Cave. Shown here is Echo Hall in 1967, prior to removal of the posts and guide wire which at one time were installed throughout the cave.

LAVA RIVER CAVE

Lava River Cave is the longest continuous* lavatube in Oregon, and one of the state's three commercial caves. Its entrance is a few hundred feet east of Highway 97, 12 miles south of Bend (1.2 miles south of Lava Lands Visitor Center). The area surrounding the entrance is a U.S. Forest Service day-use recreational site open during the tourist season. Snow on the roads and ice in the cave make winter operation impractical and risky.

There is a trail from the visitor center to the cave's end, a distance of 5,400 feet if one chooses not to visit the excavated tunnel (an additional 400 feet of stoopway). The trail from the visitor center to the actual entrance (about 400 feet) is a pleasant asphalt path through the tree-studded entrance sink. The next 600 feet, now inside the cave, is part trail, part stairway, zigzagging up and down over massive mounds of breakdown. Gradually, breakdown lessens and walking becomes easy over smooth, sandy floors occasionally interrupted by sections of moderately smooth original lava floor. At the 1,500-foot marker, where the cave crosses under Hwy. 97, the roof is about 45 feet thick. At its largest cross-section, a few hundred feet north of the entrance, it measures about 55 feet in diameter. The typical cross-section is somewhat "skull" shaped as is often the case in lavatubes with many linings.

Lava River displays most of the ordinary features of lavatubes: glazed walls; rippled floors (where exposed); breakdown ranging from slabs which slid down still-plastic walls, never quite breaking free, to massive collapse which

*Actually, there are two "caves" which share a common entrance sink. One trends northwest for more than a mile, and is the developed segment known as Lava River Cave. The other leads southeast about 1,500 feet, is largely unstable and so broken down that few original features remain. It is closed to all entry.

Much of Lava River Cave, like Skeleton Cave, has smooth sandy floors.

created the entrance; lava stalactites; secondary mineralization; and a wide range of flow marks. Sand fills the northwest (downtube) end and blankets the floor in several areas to depths of 10 feet. The sand (mostly airborn volcanic ash originally deposited on the surface) was washed in through cracks by water–probably during cataclysmic meteorological events. After deposition inside the cave, it was carved (eroded) by dripping water to form miniature badlands, some with "canyons" over ten feet deep. Today, because of carelessness and vandalism, little remains of these once magnificent sand gardens. Several attempts have been made to reconstruct them by diverting traffic to one side of the passage. The latest, a fence enclosing the most promising sand garden, appears to be at least partially successful.

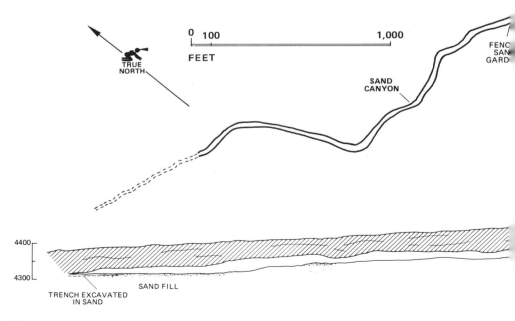

HISTORY

Lava River Cave was discovered by Leander Dillman about 1889. Dillman was a stockman and trapper who lived nearby and used the cave to cool his venison—which probably explains the many horns and bones reported just inside the entrance in 1909. For 32 years, the cave was known as Dillman's Cave, but in 1921, Dillman was convicted of a crime involving "gross immorality" and the name was changed to Lava River Cave, the name given it by Ira A. Williams in *The Lava River Tunnel*, published in 1923. This work included the first known map of the cave; a cross-section. Though not the first to clearly understand the cave's origin, Williams' perception and description of it are unassailable even today.

Williams commented about how the great natural beauty of the area around the cave entrance was threatened by logging. In 1926, possibly as a result of his comments, the Shevlin-Hixon Lumber Co. deeded to the State of Oregon, 22.5 acres of land surrounding the entrance, for a state park. In 1981, the Forest Service acquired the cave in a land exchange and continues its management as a day-use commercial cave/recreational site.

During the depression (early 1935), workers of SERA (State Emergency Relief Agency) constructed steps and a four-foot-wide trail through the entrance breakdown, shored up some threatening ceiling rocks, set posts and strung about a mile of tourist guidewire, and excavated a bypass trail in the Sand Garden area. In 1936, E.A. Zimmerman and Ray Harris spent two months digging a tunnel in the sand at the northwest end of the cave. It was determined that the sand was water-born and digging was discontinued when a sharp drop in the ceiling was encountered.

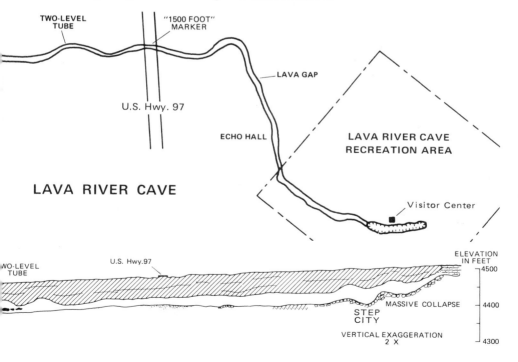

Largest of the Redmond Caves.

REDMOND CAVES

Redmond Caves appear on nearly all maps of the area (east of Highway 97 about 1/3 mile, west southwest of the Redmond Airport); appear on the earliest lists of Oregon caves, and are very well known locally. Despite all the exposure, they have never achieved much popularity with today's cavers. This lack of attention probably stems from the fact that they are, unlike many of the nearby Bend area caves, relatively small, dusty and smelly, and not easily found because of a bewildering tangle of roads and trails and lack of prominent landmarks. Once found, they just don't measure up to most central Oregon caves.

Nevertheless, they were perhaps among Oregon's best known caves in the past. Human use of them dates back thousands of years. Archaeological evidence strongly suggests they were occupied by early Indians for extended periods — probably because of their proximity to the Deschutes River. One exceptional amateur excavation yielded arrowheads, chips of many different glass-like minerals, bone awls, bone needles, bone fishhooks, scraps of woven juniper bark, a very small human skull, beads, spiral shells, mussell shells, carved ornaments, rock tools, various weapons and parts of weapons, and a tooth, apparently human.

This excavation was for several years the chief spare time occupation of four local young men during the early to mid-'40s. Nearly all of the artifacts, bones and charcoal were found in areas lighted by the entrance and at a consistent depth of two to three and one-half feet, the latter suggesting that the episode of human occupancy was, if not preceded, at least terminated by a geologic event. This large collection was in the custody of Dr. Luther

Cressman of the University of Oregon (who apparently never actually visited the site) for several years. For lack of funds, carbon dating was not performed, but Cressman comparison dated several items as " . . . at least 2,000 years old . . . " and one, an L-shaped scapula awl, at possibly 4,000 to 5,000 years of age.

The exact number of Redmond Caves depends on the explorer's definition of a cave, ingenuity, and appetite for digging in smelly dust. An early explorer recalls tunneling westward into passage where he was terrified by the rumble of a passing train (overhead). This "treacherous underground serpentine that wandered off beneath the railroad tracks — maybe under Redmond," was blocked off, according to a 1978 account. In cross-section, the caves are relatively wide. The roofs are generally 2 to 6 feet thick; ceiling heights typically range from 10 feet down to 5 feet and lower near passage ends, but are measured from the fill (of undetermined depth) that hides all original floors or breakdown. The fill is a remarkably disagreeable mixture of ash and pumice from the surface, dusty clay minerals from the inside weathering and the litter of thousands of years of visitation and occupation by humans and the ubiquitous packrats; all churned together by pothunters during the past half century. Here and there various pieces of durable litter project from the dust — too resistant to be digested by the mixture. Among these is a turn-of-the-century hot water tank — probably a remnant of a distillery.

LIONS' CAVE (One of the Redmond Caves)

A flurry of publicity, in Portland newspapers and elsewhere, heralded the opening of Lions' Cave in 1954. Named after the Redmond Lions Club, whose members opened the entrance, it was found to contain many artifacts and remains of past human occupancy. When the diggers initially penetrated the entrance, they were astonished to find footprints on the dusty floor inside. And not bare footprints, but the prints of definitely non-prehistoric sneakers!! The mystery was soon solved when two local youths admitted wriggling through the partially excavated entrance the night before.

Lions' Cave

It is interesting that proposed uses for these caves almost outnumber the actual uses. Over the past 50 years or so, various prominent people in Redmond have proposed a number of uses, none of which has materialized. Most proposals envisioned some sort of recreational use like commercialization, or creating a city park. The literature includes many references to "Redmond Cave Park." An archaeological preserve was once proposed, which was probably soon seen as closing the barn after the horse was out. One entrepeneur's proposal to grow mushrooms in the caves was rejected. As recently as 1978, the Deschutes County Historical Society was considering going underground with its museum and the Redmond Caves were on the list of potential sites. The present use of these caves, for recreation, has come about sort of by default, but is probably the highest, the best, and the hardest to prevent under present day circumstances.

Sawyers Cave No. 3.

SAWYERS CAVES

Sawyers Caves are a group of relatively short lavatubes which developed in the northern end of the Sand Mountain Lava Field. Three principal caves and several smaller examples are known in the area, and undoubtedly more exist, obscured by the area's relatively heavy forestation. The cave area is part of an east-to-west section of the lava flow. The flow probably emanated directly from Nash Crater, though possibly from a large lava lake centered near the Santiam Junction. The lake drained to the west from beneath a hardened crust. Educated guesses, plus carbon dating of wood submerged in nearby Clear Lake, place the area's most recent vulcanism around 1,000 B.C.

Cave No. 1 is entered down a breakdown slope leading to a fairly large room with a skylight. Entrances to two small, short side passages intersect the entrance room. Breakdown has created a short, low section of passage about 50 feet from the entrance beyond which little breakdown has occurred and where original floor may be seen. Cave No. 1 continues for an unknown, though probably short, distance beyond an impassably low ceiling about 450 feet from the entrance. Other interesting features are a smooth dome, beyond the breakdown pile, and a large side pocket and drag marks of a considerable subsidence near the entrance. Both Cave No. 1 and Cave No. 2 (about 250 feet southwest of Cave No. 1) appear to be uncollapsed segments of a common flow channel. Cave No. 2, much smaller, ends in breakdown.

Cave No. 3 is the largest of the group and does not appear to be related to the others. Its original floor is everywhere obscured by breakdown and rubble. Much of the breakdown and the frequent, often massive spalls and

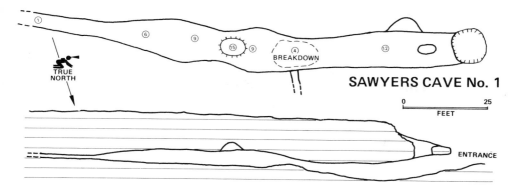

SAWYERS CAVE No. 1

TRUE NORTH

BREAKDOWN

0   25
FEET

ENTRANCE

lining ruptures occurred while parts of the interior were still plastic. Without doubt, much of the breakdown, especially that adjacent to the drip lines* is the result of frost wedging. The large entrance sinks may not be collapsed sections of a former structural roof. Rather, they may be short sections of open trench, never roofed over, for many overflows are evident along the walls near the northern entrance. Both ends of Cave No. 3 end in rubble: the southern end at the base of a lava ridge, the downslope end pinches off when floor rubble meets ceiling.

"Sawyers Cave" is located on many maps of the High Cascades and on most highway maps. Perhaps the most authoritive of these is the U.S.G.S. topographic map entitled, "Three Fingered Jack," which shows the cave's location as near U.S. Highway 20 about 2.2 miles west of Santiam Junction. The actual location is about 1.8 miles west of the junction. There is ample parking space near the caves and there may or may not be signs and a litter barrel, depending on periodic assaults by vandals. One such sign reads "Ice Caves" and while some spectacular ice forms in the caves every winter, it is not known to remain year around. There are many more small, so far unnamed caves in the area.

These caves, being among the westernmost lavatubes in Oregon and relatively near several population centers, are heavily visited. They are advertised as tourist attractions in several widely circulated publications — for example, *Sunset* magazine, which greatly contributes to their popularity. Cave No. 3 does not show the wear and tear evident in the others, despite the presence of a well worn trail leading in its direction.

*See Glossary

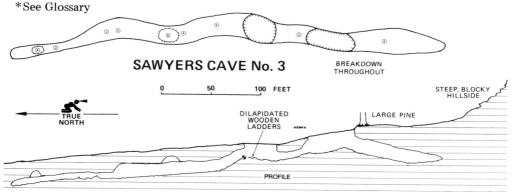

SAWYERS CAVE No. 3

BREAKDOWN THROUGHOUT

0   50   100   FEET

TRUE NORTH

STEEP, BLOCKY HILLSIDE

DILAPIDATED WOODEN LADDERS

LARGE PINE

PROFILE

35

Sandy floors make walking easy in Skeleton Cave

SKELETON CAVE

Skeleton Cave is the most popular year-around cave in the Bend area. Accessible over all-weather roads, it provides for an outing and a pleasant, sheltered hike regardless of the weather. As with nearly all lavatubes, entrance is through a roof collapse. The short upslope end of the cave (behind the stairway) is extensively collapsed causing its floor to be about 10 feet higher than the floor of the main cave.

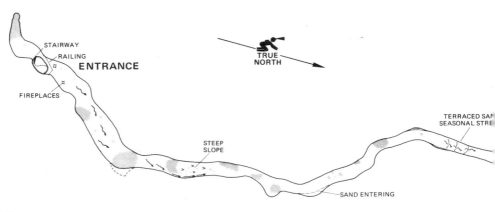

Beginning with the sturdy entrance stairway, it is easy going almost all the way to the end. Sand completely covers the floors from the entrance past the Junction Room making walking easy. Except for the last few hundred feet of the cave, beyond the lava floor tongues, the breakdown is easily avoided, and there isn't much to see except broken rock in that last few hundred feet anyway. Passages are generally wide, up to 69 feet at one point and ceilings are comfortably high in the main passage, up to 19 feet maximum. For those who don't mind stooping and crawling, there are the side passages off the Junction Room.

Except for having its sharp edges worn smooth by traffic, it is a relatively well preserved lavatube and in its features may be seen the similarities between rivers of water and rivers of lava; cut banks, meander bends and points of rapid flow. At the Steep Slope (see map), the tube angles down at 7.5 degrees, quite steep as lavatubes go, and the rounded passage shape is characteristic of rapid flow. Temperatures inside Skeleton Cave vary depending upon distance from the entrance. On a typical hot summer day, 52 degrees at 500 feet, 47 at 1,300 feet, 43 near end. During cold weather, they will range from surface temperature down to mid-30s.

HISTORY OF EXPLORATION

In early 1924, a party of Bend residents visited a cave known until then by only a few. As reported by several newspapers, they found the cave littered with bones, some so old they crumbled when touched. They also found a stick with penciled markings indicating visitation in 1894, which was, then and now, one of the earliest known records of cave exploration in Oregon.

It is probable that Phil Brogan (later known as the Father of Oregon Speleology) was a member of that party, for in later years he recalled joining a 1924 expedition to the cave, at which time "...the entrance was ordorous with a pile of must from a moonshiner's still just inside the entrance." Though the cave was known to some as Fossil Bone Cave, it was the name "Skeleton Cave," coined in a 1926 newspaper story by Brogan which stuck.

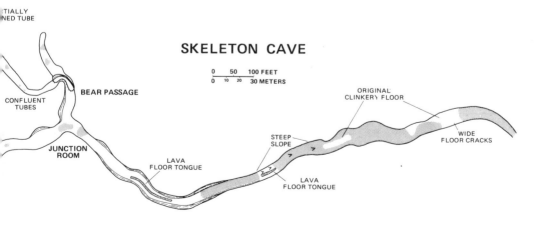

SKELETON CAVE

0   50   100 FEET
0   10   20   30 METERS

TIALLY
NED TUBE

CONFLUENT
TUBES

BEAR PASSAGE

JUNCTION
ROOM

LAVA
FLOOR TONGUE

STEEP
SLOPE

LAVA
FLOOR TONGUE

ORIGINAL
CLINKERY FLOOR

WIDE
FLOOR CRACKS

In 1928 Walter J. Perry, a government forester became interested in the cave. He is reported to have mapped it, found the main passage to be 3,036 feet, and the side passage to be 700 feet, a fairly accurate result. Unfortunately, the location of this map is not known. Perry was so intrigued by the many bones that he sent samples to the Smithsonian Institution. J.W. Gidley of the Smithsonian identified teeth found in the cave as those of a Pleistocene bear one-third larger than any living species, and in 1929, he traveled west to meet with Perry to examine the cave.

Even though the cave was by then locally well known, plenty of bones remained for examination. The remains of an extinct horse, a large hyena-like dog and the large bear, all of Pleistocene age, were inventoried. In addition, there were the bones of deer (near the end), modern horse (near the entrance) and a small fox never before known to have ranged in Oregon. Brogan had indeed named the cave well. There was much speculation at the time over how so many bones accumulated in the cave. It was thought that perhaps they had somehow rattled down through the contraction cracks in the ceiling. We cannot know, of course, whether the entrance was a natural trap for non-climbing animals during Pleistocene times, but no matter, for most of the bones could easily have been the remains of the prey of predatory animals (bobcats, coyotes and bears) which we now know can and do negotiate entrances like that of Skeleton Cave with ease.

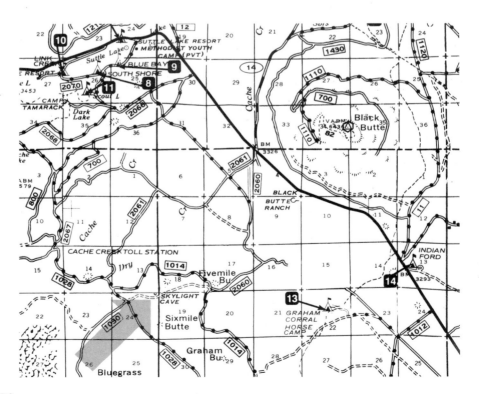

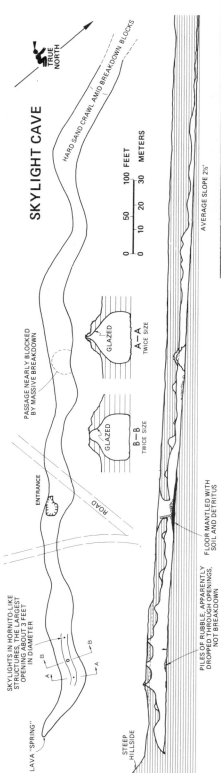

SKYLIGHT CAVE

SKYLIGHT CAVE

Skylight Cave is located about 5 miles (as the crow flies) west of Indian Ford Crossing, Santiam Highway (see sketch). Its most interesting features are the skylights after which it is named. They are unique and impart a cathedral-like character to that end of the cave. Actually, they are a combination of spatter rampart with hornitos formed along the top of the tube when it was completely full of lava, and just as interesting on the surface as inside. The cave's entrance overhangs all around and the rotting remains of many years' worth of ladders are strewn around the entrance floor, even though the entrance is not too difficult without a ladder.

The existence of many more hornitos earlier in this lavatube's active life is suggested by the many small cupolas of similar shape in the profile. Indeed, the entrance may initially have been one which collapsed. The cave's upper end is mostly original but the lower end has partially filled with sand and clay, and is punctuated full length with protruding blocks and piles of breakdown. Skylight Cave is well known, appears on most maps of the area, and obviously receives much visitation.

39

SOUTH ICE CAVE

South Ice Cave is a medium to large lava cave; medium length and large cross-section, located about six miles southeast of Newberry Crater, at an elevation of 5,020 feet, surrounded by an open pine forest. It is managed by the Forest Service as a recreation site and several improvements have been provided. Very little of the original cave remains intact. Nearly all its present boundaries are the result of breakdown — which accounts for the odd shape of the northeast passage.

A rather large complete collapse divides the remaining cave into northwest and southeast segments, 350 and 650 feet long, respectively. Locally massive breakdown further divides the cave into several large, not-too-distinct chambers — as few as three by some accounts, and as many as six by others. No branches are known except for small, buried surface tubes, one exposed just inside the southeast entrance and several on the ceiling of the far northwest passage.

The only easily recognized original lavatube features are the wall around the ice pond at Station 3N, and in the floor at the far southeastern end. From a little way up cave from Sta. 13 to the end, an Aa-aa floor is evident and overhead are angular recesses vacated by breakdown and masked by remelt. Breakdown began while the original tube was partially plastic and has probably continued at a fairly steady rate to this day and will continue. No doubt ice wedging plays a prominent role. Continuity of breakdown is further illustrated by pieces of rock seemingly "afloat" in the old perennial ice — ice which was deposited incrementally over many years. It is likely that collapse will create another entrance in the near future, geologically speaking, probably at or near Station 4.

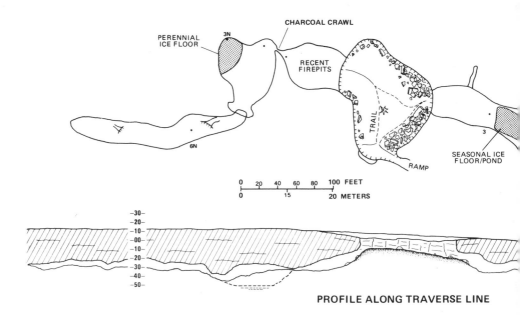

PROFILE ALONG TRAVERSE LINE

Best time to see ice in South Ice Cave is late April to early May.

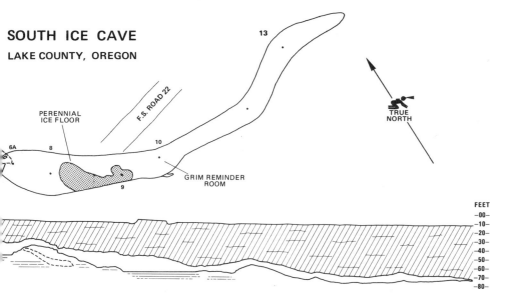

SOUTH ICE CAVE

LAKE COUNTY, OREGON

13

F.S. ROAD 22

TRUE
NORTH

PERENNIAL
ICE FLOOR

6A   8   10

GRIM REMINDER
ROOM

9

FEET
—00—
—10—
—20—
—30—
—40—
—50—
—60—
—70—
—80—

Surveyors Ice Cave.

SURVEYORS ICE CAVE

Being called an "ice" cave isn't particularly distinguishing in the Bend area where nearly all caves harbor ice at some time each year. However, Surveyor's Ice Cave contains considerable ice; probably year around. There is no record of ice there during September or November (the crucial time of year for cave ice), but it has been observed there in mid-July, and most of the prerequisites for year around ice are present.

Surveyors is located on the flank of a small, lightly forested volcanic cone on the south flank of Newberry Crater at an elevation of 5,950 feet (higher than the average central Oregon ice cave). **NOTE:** The cave is incorrectly located on most maps. The entrance is about 100 feet southwest of Road 2225 (under which it runs), *between* Roads 2225 and 2262. A well worn trail connects the entrance to either road.

This cave is a relatively short, roofed-over lava conduit (about 210 feet, map length) deeply buried by multiple flows or surges, probably from the same vent. The entrance is at the base of an 8-10 foot vertical exposure of layered lava. The southwest-facing entrance handicaps preservation of ice, but is more than overcome by surface vegetation, heavy overburden and especially the large, fan shaped entrance terrace, a highly efficient cold air collector. Access to inner sections of the cave may be difficult or impossible, depending on seasonal ice accumulation.

GLOSSARY

AA. A Hawaiian term for a lava flow with an extremely rough, jagged, spinose, clinkery, and generally irregular surface. Fully developed aa is unusual inside lava tubes. Pronounced ah ah, as in father. In Hawaiian, an expletive of pain when walking barefoot on such lava.

BENCH. A bank along the side of a lava tube. Distinction between benches, shelves and levees isn't always clear. In general, benches join both floor and wall; levees are attached to the floor and separated from the wall; shelves are attached to the wall and overhang.

BREAKDOWN. A general term for broken pieces of a lava tube's roof or walls; applied to individual blocks, accumulations, and various structures resulting from reincorporation of loose pieces in fluid lava.

CAULIFLOWER AA. Lava that has nearly completed the transition between pahoehoe and aa, the surface of which consists of closely-spaced lumps that range from about 2 to 10 inches across, that are **firmly bonded** to the underlying lava. Cauliflower aa is quite common on the floors of lava tubes. It is frequently contoured on a broad scale with billows and ropes.

CAVE. A naturally occurring void, cavity, recess, or system of interconnected passages which occurs beneath the surface of the earth or within a cliff or ledge, and is large enough to permit an individual to enter.

CAVER. One who explores caves. See also: Spelunker.

CAVING. To enter and explore caves.

CHANNEL. A long, open trough in a lava flow that carries a river of lava to a flow front. Channels inside lava tubes are typically much smaller and usually follow the tube centerline.

COLLAPSE. The failure of parts of a lava tube to withstand gravity. Collapse may occur while lava is flowing (primary collapse), or after flow has ceased (secondary collapse). Ultimately, collapse destroys most lava tubes.

COLLAPSE SINK. An essentially circular surface depression created by collapse of a lava tube roof.

COLLAPSE TRENCH. An elongate surface depression created by collapse of a lava tube roof.

CORALLOID. A type of Speleothem.

CUPOLA. A recess in the ceiling of a lava tube.

DRIP LINE. The line defined on a cave floor at the entrance by surface water dripping from the overhanging rock.

FESTOONS. Wrinkles in a thin skin of lava that have the appearance of hanging between two points. Festoons are common on lava tube walls.

FORMATION. A geological term for a fundamental unit of bedrock. It has also been indiscriminately applied to many of the fascinating features of caves, and is essentially meaningless in that regard.

GLACIERE. A cave, in rock, that contains ice.

HORNITO. A spatter cone on the roof of a lava tube. See Skylight Cave

ICE CAVE. Caves containing ice are commonly, but erroneously, called "ice caves." True ice caves are caves in ice, just as limestone caves are caves in limestone. See Glaciere.

LAVA. A general term for a molten extrusive, most commonly applied to surface flows from a volcanic vent; also, for the volcanic rock that solidifies from it.

LAVA CAVE. A lava cave is **any** cave in lava; not just a lava tube.

LAVACICLE. A general term that has been applied to a wide range of lava tube features; most often to stalactites.

LAVA FLOWSTONE. A thin fluid layer of lava on ceiling and walls in an active lava tube. Also, a general term for lava forms resulting from its flow. Compare with dripstone.

LAVA SEAL. A point where a lava tube is completely blocked by congealed lava.

LAVA TUBE. A conduit formed of hardened lava, on or within a lava flow, through which lava flows to an advancing flow front; also, a cavernous segment of the conduit remaining after flow ceases. A lava tube (while active) is continuous from one end to the other, but rarely remains that way when lava flow stops. For example, some lava tubes simply don't drain when lava flow stops.

LAVA TUBE CAVE. (Or simply "lava tube.") A specific lava tube, or segment of a lava tube that drained, and is large enough to be entered by a person. After lava stops flowing, lava tubes are left as a series of tube segments, separated by impassable areas or collapses. All parts of a segment that may be visited by a person, without passing through a collapse that is longer than deep, make up a single cave. If a collapse is longer than deep, the tube is severed and two caves exist.

LAVA TUBE SLIME. A thin layer of moist, algae-like, gelatinous mold that locally coats the walls and ceilings of humid lava tubes. Limited studies indicate that a major component is bacteria of one sort or another which account for the wide range of colors reported.

LAVA TUBE SYSTEM. A distributive network of lava tubes of the same age. A characteristic of tube-fed pahoehoe flows, and the principal means by which such flows are so widely and thinly spread. Systems are usually tree-like (dendritic) in pattern, with an identifiable trunk (the master tube).

LINING. A layer of hardened lava left against the interior surface of a lava tube by intermittent flow. See pp. 10.

LINING RUPTURE. A shallow recess formerly occupied by a thin patch of lining blown away by gas pressure, or so weakened it could no longer withstand gravity.

LOWER LEVEL ROOF. A partition dividing a lava tube horizontally into more than one level.

MAP LENGTH. The length of a cave "as the crow flies." The straight line distance between its ends (extremes). Compare with traverse length.

MASTER TUBE. The dominant tube in a lava tube system.

PAHOEHOE. A Hawaiian term for basaltic lava flows typified by a smooth, billowy, or ropy exterior and internally by lava tubes. Pronounced PAH-hoy-hoy. Literally "smooth" in Hawaiian.

PERENNIAL CAVE ICE. Ice formed in a cave, that endures from one year to the next, receding or accumulating a little each year over a period of years. See South Ice Cave.

PRIMARY. An adjective denoting events, conditions or features of a lava tube while lava was still flowing. Compare with secondary. Syn. original.

RAFTED BREAKDOWN. Single pieces or accumulations of solidified lava floated in a lava stream. Although solid basalt is slightly denser than the liquid, much breakdown floats because it contains bubbles.

ROOF. The basalt strata overlying a lava tube, usually including the initial roof crust.

ROPY LAVA. A lava flow with a corrugated surface resembling coils of rope.

SECONDARY. An adjective denoting modifications or additions to a lava tube after lava stopped flowing. Compare with Primary.

SKYLIGHT. An opening in the roof of a cave that admits daylight. A skylight may also be an entrance, but is not considered to segment a lava tube.

SLUMP BLOCK. A large block of basalt that slumps, more or less as a unit, into a collapse trench as a result of being undermined by collapse of a lava tube.

SPATTER. Small fragments or clots of violently ejected lava, commonly agglutinated (stuck together) upon coming to rest.

SPATTER CONE. A steep-sided, cone-shaped mass of spatter built up on a fissure or vent. Compare with Hornito.

SPELEOTHEM. A mineral deposit left in caves by evaporation of mineral-laden groundwater. Coined in 1952 from the Greek "spelaion" (cave) plus "thema" (deposit). A variety of speleothems occurs in lava tubes, but only two types are found in abundance: coatings and coralloids.

Coatings are thin films deposited on ceiling and wall linings, usually in bands 2 or 3 inches wide along contraction cracks.

Coralloids are tiny, coral-like projections that range in shape from acicular (needle-like) to botryoidal (shaped like tiny grapes). Both coatings and coralloids are usually light colored and stand out against the dark lava tube interiors.

SPELUNKER. One who makes a hobby of exploring and studying caves. Elitists reserve this term for recreational caving, as contrasted with "scientific" caving. Coined in the mid-30s, from the Latin root **spelunka** (cave).

STALACTITE. A cylindrical or tapering object that hangs from a ceiling or overhanging surface. From the Greek meaning "oozing out in drops." Stalactites form in all types of caves, as well as mines, vugs, veins, tunnels, hot springs, under bridges, et al. Stalactites may be composed of lava, minerals, and many other substances. Stalactites formed by the hardening of lava, also known as "lavacicles," are common in lava tubes.

STALAGMITE. A deposit on a cave floor or ledge, formed by accumulation of material that dripped from above. From the Greek meaning "that which drops." In solution caves, stalagmites are usually associated with a corresponding stalactite. However, in lava tubes, lava stalactites vastly outnumber stalagmites because the dripping material usually falls onto a molten, still-moving floor, and is carried away. On the other hand, ice stalagmites outnumber ice stalactites in lava tubes because the coldest air concentrates along the floor while warmer air rises to the ceiling. Mineral stalactites, deposited by evaporating ground water are unusual in lava tubes.

SURFACE TUBE. See page 11.

TRAVERSE LENGTH. The length of the traverse on which a cave map is based. Roughly, the total distance one would travel if visiting all the cave's passages. Compare with map length.

TUMULUS. A swelling or raising of the crusted surface of a lava flow, caused by hydrostatic pressure of underlying fluid lava.

LIKE TO KNOW MORE?

There are local caving clubs all over the U. S., most of which are chapters of the National Speleological Society, a 50-year-old nationwide organization of cavers known throughout the world as the NSS. Membership in these groups requires minimal involvement and offers many advantages like exchange of information about caves, advice on techniques and safety, cooperative purchase of caving gear, etc. If you'd like more information, contact: Oregon Grotto of the National Speleological Society, P.O. Box 2636, Vancouver, Washington, 98668.

SUGGESTED ADDITIONAL READING

William R. Halliday. 1983. *Ape Cave*. Vancouver, Washington, ABC Publishing, 24 pp.

William R. Halliday. 1976. *American Caves and Caving*. New York, Harper & Row, Publishers, 432 pp.

William R. Halliday. 1966. *Depths of the Earth*. New York, Harper & Row, Publishers, 398 pp.

William R. Halliday. 1963. *Caves of Washington*. Vancouver, Wash., ABC Publishing, 132 pp.

Charlie & Jo Larson. 1987. *Central Oregon Caves*. Vancouver, Washington, ABC Publishing, 44 pp.

Charlie & Jo Larson. 1987. *Lava River Cave*. Vancouver, Washington, ABC Publishing, 24 pp.

Charles V. Larson. *Glossary of Vulcanospeleology*. Western Speleological Survey, Vancouver, Washington, 65 pp.

TECHNICAL REFERENCES

William R. Halliday, editor. 1972. *Selected caves of the Pacific Northwest: Guidebook of the 1972 NSS Convention*, 75 pp.

William R. Halliday, editor. 1976. *Proceedings of the International Symposium on Vulcanospeleology and its extraterrestrial applications*. Western Speleological Survey, 85 pp.

William R. Halliday, editor. 1991. *Proceedings of the Third International Symposium on Vulcanospeleology*. Western Speleological Survey, in press.

Jack R. Hyde and Ronald Greeley. 1972. *Geological Field Trip Guide, Mount St. Helens Lava Tubes, Washington*. Oregon Department of Geology and Mineral Industries, Bulletin 77, pp. 183-206.

Charles V. Larson, editor. 1982. *An Introduction to Caves of the Bend area: Guidebook of the 1982 NSS Convention*. 74 pp